S0-CBG-126

The Eddington Memorial Lectures

THE CHALLENGE OF THE THIRD WORLD

THE CHALLENGE OF THE
THIRD WORLD

SIR JOSEPH HUTCHINSON
C.M.G., SC.D., F.R.S.

Drapers Professor of Agriculture Emeritus and
Fellow of St John's College, Cambridge

*The Eddington Memorial Lectures
delivered at
Cambridge University
November 1974*

CAMBRIDGE UNIVERSITY PRESS
Cambridge
London · New York · Melbourne

Published by the Syndics of the Cambridge University Press
The Pitt Building, Trumpington Street, Cambridge CB2 1RP
Bentley House, 200 Euston Road, London NW1 2DB
32 East 57th Street, New York, NY 10022, USA
296 Beaconsfield Parade, Middle Park, Melbourne 3206,
Australia

Hardcovers ISBN: 0 521 20853 X
Paperback ISBN: 0 521 09996 X

First published 1975

Composition by Linocomp Ltd, Marcham, Oxfordshire

Printed in Great Britain
at the University Printing House, Cambridge
(Euan Phillips, University Printer)

1852556

THE EDDINGTON LECTURESHIP

Sir Arthur Stanley Eddington, O.M., F.R.S., Plumian Professor of Astronomy at Cambridge 1913–44 was one of the greatest astronomer–mathematicians of his day. He was not only world famous as an astronomer but also as a brilliant exponent of the new developments in physics and cosmology. Two of his best-known books, *Stars and Atoms* and *The Nature of the Physical World,* were, between them, translated into twelve different languages. He was also a profound thinker; in religion and ethics as in science. His Swarthmore Lecture, *Science and the Unseen World* was deservedly one of the most valued and widely read of the series. It was produced in French, German, Danish and Dutch editions.

Eddington was a life-long Quaker; and on his death the Society of Friends, in order to provide for an annual lecture in his memory, established (as the result of a widely supported appeal for funds) a Trust with four Trustees; one each to be appointed by the Royal Society and Trinity College, Cambridge (of which Eddington was a Fellow for thirty-seven years) and two by the Society of Friends.

The scope of the lectureship, which has remained unchanged since the foundation in 1947, is as follows:

The lectures are to deal with some aspect of contemporary scientific thought considered in its bearing on the philosophy of religion or on ethics. It is hoped that they will thus help to maintain and further Eddington's concern for relating the scientific, the philosophical and the religious methods of seeking truth and will be a means of developing that insight into the unity underlying these different methods which was his characteristic aim.

Man's rapidly increasing control over natural forces holds out prospects of material achievements that are dazzling; but unless this increased control of material power can be matched by a great moral and spiritual advance, it threatens the catastrophic breakdown of human civilisation. Consequently, the need was never so urgent as now for a synthesis of the kind of understanding to be gained through various ways – scientific, philosophical and religious – of seeking truth.

During the period 1947–72 twenty-five annual lectures were delivered and published. Of these fifteen were given by scientists, seven by philosophers and three by theologians.

However, in 1972 a crisis arose as printing costs had risen to the point at which the advertising, sale and distribution of a single lecture pamphlet annually involved overhead costs unacceptable in relation to the price which can be charged for such a pamphlet. Indeed this crisis would have struck us many years earlier had it not been for the continued help and generosity of the Syndics of the Cambridge University Press, which we here gratefully acknowledge.

The Trustees accordingly decided, as a plan more suited to the times, that instead of having an annual lecture we should appoint a lecturer, or lecturers, every second or third year to give three or more lectures appropriate for production in book form.

This book is the first to appear under the new scheme and we hope that in this way the Eddington Trust will be able, for many years to come, to fulfil its duties even more effectively than hitherto. We are confident that it is of the great

The Eddington Lectureship

worth and distinction required for the opening of this new series and we are most grateful to our Friend, Professor Sir Joseph Hutchinson, F.R.S., for taking on, at no small inconvenience to himself, this very important task.

William H. Thorpe
Chairman of the Eddington Trustees

Jesus College, Cambridge
20 December 1974

CONTENTS

1   Introduction   *page* 1

2   The present situation   3

3   The limitations of our present system   11

4   A biologist's view of economic development   19

5   The concept of justice in society   23

6   Other communities: other practices   32

7   What kind of world do we want?   43

8   A beacon by which to steer a course   58

    References   67

ACKNOWLEDGMENTS

These lectures began as an address under the same title to a Norfolk and Cambridge General Meeting of the Society of Friends. George Baguley drew the attention of Meeting for Sufferings (the executive body of the Society in England) to the address, and I was asked to take my thinking further. Since then the subject has been widely discussed in the Society, and I have had the interest and support of individuals and of groups. To these I am most grateful. Some individuals I must thank by name. William Sewell has discussed the Chinese social revolution with me. Andrew Coulson has informed me on the social changes in Tanzania. I have had a long and fruitful correspondence with Erica Linton of the Quaker Centre in New Delhi, and from her I have received much important literature. To these, and to many others with whom I have discussed these matters, I offer my thanks.

J.B.H.

Introduction

I particularly appreciate the invitation of the Eddington
Trustees to give these lectures because of the influence
Arthur Stanley Eddington had on my life. I knew him
when I was an undergraduate, and when later in Trinidad
I read that he would be one of the principal speakers at
the 1928 Summer School at Woodbrooke, the Quaker
College in Selly Oak, Birmingham, I set about arranging
my first home leave so that I could attend. This enterprise had numerous far-reaching consequences. To get
the necessary extension of leave I applied for, and was
given, leave to work for a period with R. A. Fisher, then
at Rothamsted. That in itself had a lasting influence on
my thinking as a scientist. At Woodbrooke I listened
with interest and profit to another lecturer, Joseph Need-
ham, now Master of Gonville and Caius College. Most
significant of all, it was then that I met my wife, who
has been my partner in more than forty years of exploration, both of the world and of the spirit.

It was Eddington, however, that I went to hear, and
I was greatly rewarded. One does not remember the
contents of lectures and discussions after a period of
nearly half a century. Their influence, which was very
great, was on style of life, pattern of thinking, ideals and
objectives. But one happy little fragment of Eddington's
exposition I would like to pass on. He was expounding
the concept of time as a dimension. 'Most of you', he
said, 'think of time in terms of hours and minutes. Just

consider how limited would be your concept of a pig if you never thought of it except in terms of rashers of bacon.' This was Eddington, a man who combined superb insight with down-to-earth conversation.

Before writing this foreword to my lectures I re-read Eddington's *'Science and the Unseen World.'* This was his Swarthmore lecture to the Society of Friends, delivered in May 1929. The thoughts there expressed must have been very much in his mind when he talked to us at Woodbrooke in August 1928. As I read, I came to this: 'The essence of the difference [from the laws of science], which we meet in entering the realm of spirit and mind, seems to hang round the word "Ought" '. The concept of 'Ought' has been on my mind ever since I began on this exercise, and what I have attempted to do is substantially to follow Eddington and first to separate out those things that we must accept as the facts of life, and then to consider what we ought to do in those other matters where value judgments are required of us.

In this essay I have trespassed in the fields of disciplines other than my own. I make no apology for doing so. Indeed, I believe that in so far as we stick conscientiously to our own specialisms, we fail to comprehend the world in which we live, and the problems we have to solve.

2

The present situation

Having spent the greater part of my working life in countries of the Third World, it is natural that my view of the world as a whole should be coloured by the experiences and problems of poor countries. The extent to which I shall discuss the Third World in these lectures is, however, due to something more than long familiarity and years of close contact. I have found that in taking what the Western World has to offer and deploying it in the Third World, one comes to see more clearly than I have been able to do in any other way, the constraints and limitations of the social and economic system that we in the West have devised. In this system the challenge of the Third World that I want to speak about is a challenge to the West, a challenge to our ideals – or lack of them – to our standards and our values, and to the ambitions and aspirations that we have for ourselves.

It is not possible to define the Third World, but it is important to give some account of its enormous diversity if we are to appreciate the nature of the challenge it makes to our own society. The three great continental areas involved – southern Asia, Africa and Latin America – differ fundamentally in their social structure and in their relationships with the rest of the world. Southern Asia is a region of ancient and highly developed civilisations with old-established cities in large and politically advanced states. Africa at the time of European penetration was a continent of tribal societies, fragmented, and denied the prospect of social advance by conflict within and by raiding from without, for cattle and for slaves. Latin America suffered the imposition of a Mediterranean civilisation, together with massive colonisation. The indigenous cultures, some of them advanced, were overwhelmed to a degree that has not happened in Asia or Africa.

These very different culture areas have in common the imposition by European powers of an economic system, based essentially on the Industrial Revolution, that emerged in Britain and spread to Western Europe and North America in the late eighteenth and the nineteenth centuries. The European economic system was established in the Third World by the introduction of a money economy linked to the European monetary system. This was a necessary step if any kind of development was to take place under the imperial and colonial powers. The establishment of an administration required the kind of division of labour that makes possible the preparation of a state budget and the imposition of taxation to finance government expenditure.

In most of Africa this was quite new. It supplanted systems of barter and of tribute in kind, or at best the use of cowrie shell currency. In Latin America, Spanish and Portuguese monetary systems were a part of the Mediterranean civilisation that supplanted local cultures. In southern Asia, monetary systems were already in operation and they were taken over progressively, and linked with the European exchange system. This was a slow process. When I was in India in the 1930s I worked in some States which still coined their own rupees based on silver, while the Government of India rupee was based on sterling.

Development followed the establishment of administration. The motive forces of the imperial and colonial era were diverse. The hope of gain was strong in some areas, notably India. Power, and in particular power against rivals among the metropolitan states, dominated others, particularly Africa at the time of its partition. But there were also altruistic motives, particularly those of Christian missions and of the governments with which they had influence. Over all there was a confidence in the superiority of Western civilisation that many of us today find it difficult to endorse. These diverse motives fell into a loose coherence through the economic philosophy of enlightened self-interest.

Thus arose the first stages of development as we understand it. First and foremost came communications. Administration was only really effective when it was possible to move freely about the territory administered. Suppression of tribal conflicts, of slave- and cattle-raiding, and of such anti-social phenomena as the Thugi murders in central India, depended on ease and speed of movement of government officials, police, and military forces. So roads were improved, lake and river navigation was established and, above all, railways were built and operated.

The other great means of communication, the written word, was for a long time almost entirely developed by Christian missions. Education in the Third World is still closely associated with the Churches. In India, state responsibility for education developed early, but many of the schools and the universities were generated by missionary enterprise, and for long were supported both in staff and in finance by missionary societies. In Africa the school system was heavily dependent on the mis-

5

sions, both Catholic and Protestant, right up until the second World War. In Uganda, for example, the Government only assumed full responsibility for the school system after the de Bunsen report of 1953. Here let me pause to pay a tribute to the late Sir Andrew Cohen, an alumnus of Cambridge University, and in a real sense the father of Ugandan independence. He laid the foundations, and built much of the superstructure, of the state education system in Uganda. He initiated technical education and he nurtured the infant Makerere College until it reached the full stature of a university.

In the communities integrated by the new communications system, general living standards rose. Personal security, initially precarious, was improved not only by the establishment of law and order, but much more by the development of public health and medical services. Where I worked in India in the 1930s, two open graves were regularly maintained in the British cemetery. This kind of forethought was by then obsolete, and the only occasion when one was required in my time was to bury the victim of a motor accident. Nevertheless, the tragic records on the gravestones bear witness to the toll of disease among young Englishmen in the early days of the British Raj. Likewise in Africa, the record of deaths among young missionaries shows the hazards to life in the early years. After the second World War these hazards were greatly mitigated by improvements that resulted from the development of medical services.

An economy linked to the world's monetary system only becomes meaningful when there are goods to be exchanged. In Latin America one of the great incentives to conquest and colonisation was the existence of gold and silver, both in current use and to be won by mining.

6

The present situation

The wealth of India was more in the produce of agriculture and of craftsmanship. Indigo dye, cotton and silk textiles, and crafts in wood, metal and precious stones were among the goods on which the fortunes of the East India Company and its servants were based. Africa had less to offer. Indeed, it required considerable effort to make Britain's African colonies even self-supporting. The Uganda railway, for example, was built from Mombasa to Kisumu to provide a line of communication for Uganda in which there was a strong missionary interest. Only after it was built was it decided that one means of generating traffic to enable it to pay was to encourage European settlement in the highlands of Kenya. In the event, the contributions of African territories to the trade on which the money economy depends, were two-fold. First, a range of crops that became known as 'cash crops' was established, which had the common characteristic that they were in demand in temperate regions but could not be grown there. Secondly, ore bodies were discovered, and mining and smelting established, and the products exported to the metropolitan countries.

The need of the Third World was for industrial goods. Everything from steel rails and locomotives to sewing machines and bicycles was on the shopping list. Indeed, much that had formerly been produced locally was available more cheaply and in greater quantity from the industrial countries. The iron smelting and hoe making crafts of equatorial Africa gave way before competition from Birmingham. On a much larger scale India, which introduced cotton to Britain by way of her hand-made textiles, was driven out of her home market by the massive competition of Lancashire's industrial textile production.

7

In this way the Third World became integrated in the trading system of the Western World. The new economy and the trade links with other parts of the world called for changes in the structure of the developing communities. Trade was dominated by the exchange of the farming products of the country for the industrial products of Europe, and later of America. Hence the trading centres that grew up were at the nodes of the overseas traffic routes. Most of the new cities of Africa are at the ports, or at the communication centres of railways and river and lake transport. They served, rather than were served by, the transport system. Even in India, the three great port cities of Calcutta, Bombay and Madras grew up to serve the international trade generated by the imperial system, rather than as integral parts of Indian social and political life.

All this was excellent in its day. Western countries had much to offer. They provided capital goods through loans and consumer goods on current account, and they made possible the building of railways and harbours, and civil engineering works such as dams, reservoirs and canals. In return, peasant farming communities as well as the business fraternities of the new cities benefited in the availability of goods and services formerly undreamed of. It would be interesting to compile a balance sheet of the exports of wheat and imports of capital goods and irrigation and engineering services through Karachi in Pakistan during the British Raj, or of the cacao exports and industrial imports through Takoradi in Ghana during the colonial period.

We do well to remember all this, and the enormous advances made in the Third World in the nineteenth and early twentieth centuries. It is easy to criticise, to point to social attitudes that we would now regard as intolerable, and to economic practices that many would consider exploitive. To do so is both unprofitable and unworthy; unprofitable because hindsight is the lowest form of wisdom and unworthy because so much of our current practice will appear suspect in the light of the understanding of a quarter of a century hence.

Having agreed that the achievements of the imperial and colonial period deserve our respect, and even our admiration, let me now acknowledge that we are today in a state of disillusion. The 25 years since 1945 may be regarded as a period of transition during which the imperial and colonial system progressively yielded to independence. It was characterised by a rapid rise in wealth in the world as a whole, a substantial part of which was in the Third World. Yet progress was uneven. The rich got richer and the poor stayed poor. What Lord Blackett (1967) called 'The ever-widening gap' developed between the rich and the poor countries, between different poor countries in the Third World, and between the urban rich and the urban and rural poor in all countries. The gap between affluent Britain and poor Tanzania, for example, is matched by that between developing Pakistan and stagnant Bangladesh, and between the industrial magnates of Delhi and Bombay and the peasants of Rajasthan and Maharashtra.

We used not to be worried by this. Freedom was freedom for the best man both to win and to hold on to his winnings, and equality was something that would be achieved when the increase in wealth of the community enabled the poor to catch up. The very pace of change and of enrichment has revealed the weaknesses in this philosophy. Enrichment has gone so disproportionately to the already rich that the gap between rich and poor widens instead of narrowing as the general level of wealth goes up. Further, it has now become apparent that the resources of the earth that are available to man are finite, and the rate at which they can be discovered and exploited is not such as to enable us to look forward to a time when the Third World could achieve the standards of resource use now customary in the West.

Thus our long-term philosophy has failed us, and we need to set our sights afresh. There is much that we can do – that we must do – but we cannot look forward to achieving even more modest goals by a continuation of the process whereby we made the first advances in the Third World.

It is the recognition of this situation that is the basis of our disillusion, but it should be said that disillusion is not the only possible response. Our forefathers devised an economic system with which they achieved great things. That these achievements are in the past and we have little expectation of further advance on this line means no more than that we need to think again. Systems are expendable. Provided our goals are not in doubt, a new system to meet the new situation can be devised.

3

The limitations of our present system

The system by which we achieved the first advance in the Third World was that of the exchange of primary products for industrial goods. It will not serve the need for further advance for a simple reason. The exchange of raw materials for industrial goods is not a sufficient condition for development to take place. The experience of the Third World in marketing its primary products has been of wide variations in price in response to changes in supply. In general the benefit to the producer of any substantial increase in supply has been greatly reduced by a fall in price. Over all, the terms of trade have been to his disadvantage. For this unsatisfactory situation there has been a tendency, among those sympathetic to the needs of the Third World, to blame the economic policies of the West. Yet the problems of disposing profitably of the products of the Third World are not confined to cane sugar, which is in direct competition with a Western product, and cotton and jute, which are in competition with man-made substitutes. They have arisen also with coffee, tea and cacao, which cannot be produced in temperate regions. In fact, the supply and demand relationship for primary products is much less elastic than it is for industrial goods. Thus, seasonal effects on the supply of cacao, for example, have enormous effects on cacao prices, just as a degree of overproduction coupled with competition from synthetic substitutes can ruin the market for natural rubber or for jute.

The Third World has consequently come up against great impediments to development by the traditional means of exchanging primary products for industrial goods. This is not just a matter of adverse terms of trade or of short-sighted or selfish protectionism by the West. Western trade policies are devised – maybe inexpertly, but nevertheless devised – as a response to economic circumstances, and it is these circumstances that must be taken into account. Restrictions on trade are symptoms, not causes of slow development.

Essentially, the growth of advanced human communities depends first on rising agricultural productivity, and secondly on the growth of crafts and industries outside agriculture. The adequacy of the food supply is crucial. If food is insufficient or even uncertain, all else must wait until it is assured. If it is adequate, further increase will be superfluous, and will be wasted or go into luxury consumption. As the production of food is increased, the possibility of developing crafts and industries rises because labour can then be spared from agriculture. Thus an urban sector of the community arises, manned by labour released from the rural sector, dependent on the rural sector for food, and supplying in return the goods and services by which man's estate is improved after his basic need for food has been met.

It is in the relationship between town and country, between industry and agriculture, between the industrial West and the Third World that our thinking has been inadequate. We have been concerned with the exchange of produce, goods and services, and we have forgotten that as change and development goes on, the balance between them is maintained by the movement of people from rural to urban communities.

12

Consider this in terms of the increasing productivity of agriculture in the Third World. In the British Industrial Revolution, labour moved from the farms to the factories in the neighbouring towns, and the proportion of the population engaged in agriculture has fallen from something of the order of 70 per cent to less than 4 per cent. As West Africa improves the production of cacao farming, for example, the same process of transfer should go on, since we do not want much more chocolate, and West Africans do want a great deal more of the goods and services of urban communities. But the cities they serve are over the ocean, and out of reach of potential emigrants. It is interesting to note that those who are alarmed at the entry of other races into this country are concerned over an almost imperceptible trickle compared with the flood that would be necessary to balance our trading relations with the Third World. For let there be no mistake: we have created a trading area in which the Third World is our rural sector and we are the urban sector. And if the trading system we have imposed on the Third World were to be made to work, it would not be so much their produce of which we should take more, but their people.

This is manifestly impossible. We have neither the room nor the resources to complete the industrial revolution by transferring the rural population of our trading area to our own cities. It is in this sense that protection, against both the import of produce and the migration of people, is a symptom of the circumstances under which we live rather than a cause of economic ill. It follows that if the Third World is to enjoy the advantages of development, the transfer of labour from agriculture to crafts and industries must go on as it went on in the industrial development of Europe and America. It must go on within the Third World, not from the Third World to the West.

This is the justification for industrial development in the Third World, and for the protection that is so generally given to their infant industries. It is beside the point to argue that these goods would be obtained more cheaply from the industrial countries. There is not the means of exchange to buy them in sufficient quantity, and there is no prospect on that basis of raising the living standards of the rural poor of the developing world.

Industrialisation has come by transfer from the Western World, and has in consequence involved the acceptance of Western standards of technology, use of capital, and conditions of employment. Since these standards have evolved over the whole period of the Industrial Revolution, characterised by rising living standards, demands for more and better goods and services, and insistence on improved working conditions and shorter hours, they have come to be highly capital intensive and labour saving. So their application to the Third World, where labour is abundant but unskilled, and capital is scarce and costly, has had little influence on urban employment but a large effect on the emergence of a small wealthy minority.

In the Third World as a whole agricultural productivity has been steadily improved throughout the present century. Increased agricultural productivity has made possible the progressive development of the services of a modern state – law and order, communications and transport, and education – and the introduction of the goods produced by industry by exchange with industrial countries. Transfer of labour from agriculture to industry

14

and trade has begun. Capital cities and trading towns and ports have served as the foci of urbanisation, and industrialisation has been initiated under the shelter afforded by policies of import substitution and protection.

Serious constraints on further development have now become apparent. The tendency to concentrate non-agricultural activities in a few rapidly growing urban centres is a copy of the pattern of urbanisation in the industrial West that was set by the need to congregate urban labour within walking distance of factories run by steam power. The more dispersed, and socially more desirable, small town industrial pattern of the mediaeval English woollen industry has been forgotten. Migration out of agriculture has become very substantial, but the demand for labour in urban activities has not kept pace with it, and there have grown up around the cities of the Third World shanty towns in which migrants from the rural areas lead poverty-stricken and unhealthy lives, unemployed or under-employed, and dependent on such support from friends, kinsmen or tribesmen as they can get. Against this background there has also developed the small minority of urban rich, not only of politicians, civil servants and industrialists, but also of those who have gained employment, and who have safeguarded their own position by organising themselves into trades unions and trading fraternities so that the limited opportunities for labour and trade are preserved for their own groups.

I set this out, not in criticism of the Third World, but to set up the Third World as a mirror of our own. Development in this century has been Western development, inspired by Western ambitions and objectives, and

15

practised according to the rules of the Western game. Moreover, the faults and fallacies in any system are best identified where the system is applied outside the circumstances in which it was generated. So with this appraisal of the state of development in the Third World let us look at our own world. From the days of the mediaeval craft guilds to modern times with employers' federations and trades unions, the motive force of development has been self-interest. The guilds reserved the practice of their craft to their members as a modern trades union reserves the right to its own work to its own members. And as there has built up a sad reservoir of underemployed in the shanty towns of Africa and Latin America, so the towns of mediaeval Britain were plagued with the 'sturdy beggars', the surplus labour released from agriculture and knocking on the closed doors of the towns. The balance between town and country has been worked out in a context of self-interest that has led to periods of unemployment on the one hand and of self-defensive organisation on the other. This is not new. It is not even recent. Two thousand years ago Jesus said, 'To him that hath shall be given, and he shall have abundance.' So the rich get richer, and since we have so passionately advocated freedom, freedom for the individual has come to mean freedom to get what you can, and to keep it for yourself. And so the poor stay poor. There is a premium on organisation, since organisation gives power. And large organisations gain greater power, so the principle of self-interest favours the large organisation. So power and wealth aggregate in large groupings, and there is no place for the small group or the individual.

We have begun to question this system. We view with satisfaction and some pride the development of the

Welfare State. We regard it as a great advance. But in fact what we have created is essentially a scaled-up model of the system of social security maintained by the extended family, and the chief and his clientele, in simpler and smaller social systems. We have undertaken some redistribution of resources so that a man shall not starve, and this is what is achieved by a tribal community at subsistence level. The basic resource is land, and it is distributed by the chief so that all members of the tribe have right of access to it.

It is perhaps an achievement of some merit to have mitigated the effects of self-interest thus far, but it is not enough. We have set out to see that a man does not starve, but we are content to let him live in beggary. The redistribution of resources is by those in power, and indeed the power of the powerful is enhanced in the process. We have not yet reached the stage of contemplating seriously the redistribution of power. And only by that could we hope to rescue men from beggary. Our whole attitude to unemployment pay and national assistance is to emphasise the obligation of the poor to the powerful. Relief must be set at such a level as to maintain the incentive to seek – even to supplicate – for a job, any job. Yet there have repeatedly been periods in which there simply were not as many jobs as there were people seeking them. At such times the pressure should surely have been on the powerful to provide the jobs, not on the poor to seek what manifestly was not there.

Let me say again, I am not accusing any one or any system of perversity or wickedness. I am pointing out what I believe are the consequences of a system based on self-interest. It is logical, and it has been used to make great changes, and in many cases great improvements, in the human situation. But it has fundamental drawbacks, in that of its very nature it creates and then exaggerates differences between one man and another that have no foundation in equity.

This is the root of our disillusion. If liberty means licence to exploit any advantage I may have over my neighbour, and to hold on to that advantage as of right, what meaning can I attach to equality, and in what sense can I claim a fraternal relationship with him?

4

A biologist's view of economic development

I am a biologist, and economic development has biological
as well as economic connotations. For this reason if for
no other I believe it is relevant to set out a biological
point of view of social progress. The development of our
present system has been a long process, but the establishment of its salient features took place during the Industrial Revolution. This was not a planned revolution. It
came about as men seized the opportunities presented by
the invention of power systems and of machines to
improve the productivity of labour. Nevertheless, it
called for explanation, and the classical economics of
the nineteenth century was worked out as a means of
accounting for what was going on. It seems to me very
important to recognise this. If – as I believe – our
economic system is inadequate for our present needs,
we have to choose between amending our needs to conform to the dictates of the system, and altering the system
to take account of our needs. If our economic system
represented an interpretation of natural law, we would
have to do the former, but if we can accept it as no more
than a good account of the economic phenomena of the
nineteenth century, we need have no hesitation about
revising it in the light of late twentieth-century experience.

Let me recount the generally accepted stages of economic development, and give a biologist's view of them.
First there are the stagnant, or very slowly growing,

social systems of communities untouched by 'development'. It has been our main preoccupation to get these economies moving, and hopefully we achieve the 'point of take-off', which is stage 2. This leads on to the third stage, which is that of self-sustaining growth. Growth was the characteristic of the Industrial Revolution for which the economists of the nineteenth century set out to account. In this century we have tried on the one hand to maintain self-sustaining growth in industrialised countries, and on the other to establish it in those parts of the world – which we call 'developing' – in which the economy is static.

These three stages may be interpreted as the first part of a relationship very widely observed in biological studies of growth, which is represented mathematically by a sigmoid curve. Growth proceeds slowly in the beginning. A point is reached where sufficient resources have been mobilised to allow a great increase in growth rate. The new material so produced contributes directly to the growth process, and the situation is achieved that the economist observes as self-sustaining growth, and that the biologist terms the 'grand period of growth'. In biological systems this phase is followed by one in which growth slows down. Increasingly, the products of growth are devoted to materials that do not contribute to further growth. A tree lays down wood, which is necessary to support the structure but does not generate growth. A wheat plant produces seed, in which stores are laid down for growth the next season, but not for the season in which they are made. So the latter part of the curve flattens off, and growth either ceases or goes on at a new rate similar to that of the first stage.

Growth in human communities is also a biological

process, and it has characteristics similar to those of growth in plants or in simple populations. The resemblance between the first three stages of economic growth and the first part of the general biological pattern of growth is clear. I believe that much of our economic trouble in this country in recent years has arisen from our failure to realise that we have reached a stage in economic growth in which much of the product of growth goes into non-growth material, and in consequence our grand period of growth is over. Consider such everyday services as the passenger transport system, the police and the hospitals. They are all inadequate to meet what we regard as our reasonable needs and they will doubtless absorb a larger share of our available resources in the future. But during the grand period of growth of the British economy, the share of our resources devoted to them was much less, and the difference went predominantly to growth activities. People had to live within walking distance of the work place, and spent a larger proportion of their time working and less in getting to and from. Such policing as was practised did not involve the diversion of men from productive work to traffic control and the management of car-parking. In the early days of the Industrial Revolution the sick and the injured got better or died, with only very limited assistance from doctors and hospitals. In all these ways we are better off, but we must recognise that in devoting the products of economic growth to these most desirable ends, we are diverting them from the function they had in the grand period of growth, of generating yet further growth. It is for reasons such as these that the slowing down of the rate of growth of an economy such as ours must be accepted as a part of the natural sequence of stages of growth.

21

The diversion of the products of growth into channels that are not themselves growth promoting is something that only the affluent can afford. We know this but we have been curiously unaware of the consequence that we can only do it at the expense of the rate of economic growth. The passenger transport system is like the wood laid down by a mature tree. It is a necessary support to the structure, but the resources put into it do not contribute to further growth. We may carry the analogy further, and regard the Welfare State and the hospital service as comparable to the seed produced by a maturing crop. They improve the prospects for coming generations but they do so at the expense of current growth.

The concept of a phase of declining growth rate following the grand period of growth provides a framework into which much of the thinking, discussion and experimentation in our economy appropriately fits. The environmentalist, exploring the idea of 'zero-growth', clearly has a contribution to make. Those who put forward proposals for a 'just society' are in fact concerned with the allocation of resources to those no-growth sectors of the community that the West can afford, though the Third World cannot.

This is development as I see it; a continuous process characterised by a slow beginning and a slow latter end. We have been so obsessed with exponential growth in its middle section that we have failed to see the process as a whole. We need now to achieve a comprehensive view if we are to steer a steady course as parts of the Third World move into the period of rapid growth on the one hand, and the West faces the slow growth of a mature social system on the other.

The concept of justice in society

In so far as a 'just society' can be defined, the conception involves the distribution of goods and services according to need. Our current system of distribution is according to wealth, through the operation of the price mechanism. Wealth generates wealth, and distribution according to wealth creates a gulf between the rich and the poor. In our pursuit of social justice, therefore, we are particularly concerned with experiments in balancing supply and demand other than by wealth. This, of course, will not be our sole concern. We shall need also to devise criteria by which we determine the allocation of resources between the growth and the non-growth sectors of the economy. But of first importance are those attempts at distribution according to need that can serve as guides in an attempt to move towards a greater measure of social justice.

I say to move towards social justice rather than to establish a just system, first because we have as yet no clear concept of a just system and we shall only develop one as we work it out in practice. A second reason is almost equally important. Much of the current advocacy of change is concerned with sweeping away what we now have. To get rid of the old and to set out to rebuild on a clear site is an attractive proposition. Even in a simple subsistence society it would not be a practical one. A man must eat day by day or he cannot survive, let alone build a new community. In an advanced society, with a tremendously elaborate division of labour, and with urban communities of immense complexity, the rate of change must only be such as will permit of the maintenance of the life of the community through the transitional period. Indeed, we are even now experiencing a rate of change such that we are in danger of suffering the breakdown of services and the consequent advent of famine and disease in our big cities. On both counts it is vital that we work out a new system without at any stage jeopardising the old whereby we now live. We must take advantage of every experimental approach there has been to the management both of production and of distribution according to need, and use such information to devise means whereby we can further shift the emphasis of demand from wealth to need.

For examples I naturally turn to agriculture, partly because of my familiarity with it, but also because agriculture provides the best examples of attempts to manage the economic process. It employs the basic natural resource – land – and provides for the basic need – food – of all communities. Hence the frequency with which man has experimented with market management in the agricultural world. Interference by the community in the market mechanism through which the individual does his business goes on increasingly both in the transfer of the basic resource, land, and in the distribution of the product, food.

Land is limited in amount and fixed in location, and these characteristics are reflected in the advantages of ownership. It has become clear that no community can tolerate absolute rights of ownership because community needs for land can only be met by land in particular locations. So powers of acquisition have been taken, and control of use exercised through planning authorities. These restrictions on the absolute right to property in land have been imposed as far as possible within the price structure, and compulsory acquisition is based on the theory – somewhat modified in practice by planning restraints – that it is paid for at what is estimated to be the value if it were subject to a contract between a willing buyer and a willing seller. There is no means of preventing the price of land rising on account of scarcity, and indeed a combination of scarcity and location has led to land prices in and around urban areas rising to levels that must be regarded as quite inequitable.

For much of Britain's history, individual rights in land have been of great value in the development of the economy. Considering the extent to which individuals have profited at the expense of the community in recent years, however, it must be accepted that the minor restrictions on the land market imposed by planning authority and by powers of compulsory purchase are inadequate for present circumstances, and more extensive restrictions on rights of ownership of land are necessary in the interests of social justice. Indeed it is well to remember that in many parts of the world there are no individual rights in land. Land is an asset of the community, and individual rights are limited to rights of use. If present inequities cannot otherwise be remedied, we may need to resort to such a system. We must await with interest, therefore, and indeed with sympathy, the present Government's proposals for the control of land required for urban development.

On agricultural production, the history of the supply of food for the British people is a history of management, and not of free play of the price mechanism. As Ernle (1919) has put it,

> Beginning in the early Middle Ages, and ending in 1869, the English Corn Laws lasted for upwards of six centuries. Attention has been so exclusively concentrated on one side only of their provisions, that the regulation of the inland trade in corn and the restrictions on its exportation have long been forgotten . . . The general aim of legislators was to maintain an abundant supply of food at fair and steady prices . . .

Only in the latter part of the nineteenth century, when there was a steadily rising demand for food and ample land and labour available to give a corresponding rise in supply, was the balance between supply and demand maintained by the price mechanism. Even so, the balance was short-lived. With the opening of the American West, land became so readily available that prices fell inordinately and British agriculture was reduced in size and in productivity by the long depression.

With the second World War, the attempt to leave food supply to the free play of the price mechanism finally failed. Supplies were limited and their distribution was controlled by rationing. Prices were determined so that all, or nearly all, of the community could buy the full ration. After the war, shortages persisted so long, and the interludes of plenty have been so short, that there has been little opportunity to return to the free operation of the price mechanism. Interference with the market for food is now an almost world-wide practice. After a short spell of prosperity in an expanding market, and a longer spell of depression through overproduction, free trade has given way to experiments in the management of the

26

food market akin to the practices of the six centuries prior to 1869, rather than to the *laissez faire* of the late nineteenth century. I do not believe that this is a matter of economic ignorance, or of political wickedness in high places. It is simply that the conditions under which price serves as an adequate control mechanism for the supply of food to the market are very limited. Circumstances in which it is necessary for a community to take positive steps to ensure its food supply are of much more general occurrence.

The vast expansion of supply with the opening of the American West and the disproportionate expansion of demand following the second World War were equally disastrous for control by price. Further, the two sides of the price equation are governed by different time factors. Supply is a matter of next year's cereal harvest, or of the beef that will be produced by a 15-month old steer that will not be born until 9 months hence. Demand is concerned with tomorrow morning's breakfast, and will hardly tolerate a delay of a week. The tragedy of this situation is before us at this time particularly. The nations debate the world food supply situation while we know, first that we have no reserves with which to counter famine, and secondly that even if we gather up the crumbs from the rich men's tables – our tables – the poor in Bangladesh and India will die in their thousands before our meagre aid can possibly reach them.

I have spelt this out because in the food market management by price has been ineffective, and we in Britain have had to follow our forefathers in experimenting with other means of management if we are to avoid situations in which a balance between supply and demand is achieved by the crude resort of letting the poor starve.

It is for this reason that the study of the management of food supply is particularly relevant to our problems.

Two experiments in managing supply and one in controlling demand deserve our attention. Consider first the marketing of milk in Britain. When my father sold milk, production was in the hands of a large number of farmers who were unorganised. A few well-managed dairies organised an efficient supply to the big cities, and thereby gained control of a substantial share of the market. The price to the farmer was squeezed, and he had to organise to meet the pressure. The need to protect a depressed producing industry, followed by the desperate needs of consumers during the war and post-war years, led the Government to encourage producer organisation, and milk marketing is now a monopoly in the hands of a statutory board. The price mechanism has been displaced, and the Government represents the consumer in the market bargain. Prices and supply policy are determined by agreement between the farmers' organisation, the board and the Government, with the Government in the last analysis having the final say. Milk marketing is perhaps the best and most hopeful example of the experimental development of a system of trading outside the market mechanism. The second experiment is the Price Review system whereby since the 1947 Act prices for agricultural produce have been negotiated between the Government and farmers' organisations. This has given annually an opportunity to review the needs of the community and the achievements of the industry, and to manipulate the price structure according to the Government's assessment of the needs of the country. Without entering into debate as to how correct the assessments of needs have been, there can be no doubt of the impact

28

of the Price Review system on the structure of British farming. For example, to meet wartime needs great emphasis was placed on cereal production, and this has been a feature of the price structure throughout the post-war period. The total acreages and production of the three main cereals (wheat, barley and oats) before the war, at the end of the war, and in 1970 were as follows:

Year	Acreage (millions)	Production (million tons)
1935	4.0	3.4
1945	5.5	6.9
1970	7.7	10.8

There have been very large increases both in area sown and in productivity. Moreover the system of Government backed price guarantees has been an instrument in the implementation of our developing concept of social justice. By dividing the price of many of the basic food-stuffs into a support element and a market element, the country has been able to combine the security afforded by a productive home agriculture at the expense of the taxpayer, with low food prices in the market for the benefit of all, including the poor. The impact of wealth in the food market has been reduced; perhaps only marginally, but nevertheless reduced. In this situation there is an element of redistribution of wealth in the incidence of the taxation by which the support element is financed. In circumstances such as the present, in which the supplies available to us are limited, some further restraint is required, and I suggest that this should be by rationing.

29

Food rationing during the second World War was a great and instructive experiment. By this means a strictly limited supply of food was distributed in such a way that the whole population regardless of wealth had access to specific allowances of food. These were calculated having regard to nutritional requirements, and to the special needs of particular groups such as men engaged on heavy manual work, pregnant and nursing mothers, and growing children. Rationing was not independent of the price mechanism, and price control and support from taxes were necessary to make it work, but the essence of the matter was that distribution was dominated by rationing according to need and not by price according to wealth, and thereby a much greater degree of equality was achieved between rich and poor. Wartime rationing was not, of course, just a success story. Anything improvised under such stresses would have weaknesses and failures. But it did give experience, of great value to us, of an alternative to the price mechanism, in which need was a major consideration.

These experiments with alternatives to the price mechanism were undertaken under the stimulus of great economic strain, and there was always pressure to revert to the familiar patterns of trade when conditions became easier. Rationing has gone. The Price Review will be replaced by the Common Agricultural Policy – a much less useful experiment in meddling with the price structure, if I may venture my own opinion. Only the Milk Marketing Board remains. This may seem surprising in a world that increasingly pays lip service to social justice, but it is a consequence of our failure to think out ways in which we can move from a wealth-dominated social system to one in which an obligation is accepted to distribute at least essentials according to need instead of according to ability to pay. Let me repeat: we have not done our thinking on this. Wartime controls both of production and of distribution were worked out under the pressure of a national emergency. They were, on the whole, very good, and they represent practical experience that is of great value to us, but they are not a sufficient guide on which to develop a system that will be fair, not only to consumers, but also to producers – farmers and farm workers both in this country and overseas. We have much thinking still to do.

To sum up, in this affluent world in which we of the West live, there are endless claims for even more of the goods and services to which we have grown accustomed. The extent to which we have already gone towards meeting these demands in itself reduces our ability to generate the growth by which more might be provided. Yet in addition to this we now have a growing sense of responsibility to distribute goods and services according to need. This is socially right and proper, and the existence of such a sentiment in our society is encouraging for the future. But if we are to discharge this responsibility, we must devise instruments of policy other than those appropriate when, during the grand period of growth, the products of growth were used predominantly for the generation of further growth. Rationing is such an instrument, and one which deserves close consideration free from the background of wartime extremity. Hitherto we have looked on rationing as a resort in a crisis. It now merits assessment as a policy for social progress.

6

Other communities: other practices

Western economic practice, though it has dominated the world in the twentieth century, is not the only system by which men organise their communities. Other forms of community organisation are in use, in India, Russia and China. These forms of organisation must be considered, and their relevance to the solution of problems raised by our own way of doing things assessed.

It will be helpful to put all these countries, together with the rich Western countries, into a common framework that may for convenience be called the development process. This may be seen as a progressive liberation of human communities from the need to devote the whole of their energies to meeting their subsistence needs. As a society improves its agricultural competence, so some members of it are released from the necessity to produce their own food, and devote their attention to the satisfaction of other wants.

The whole process of elaboration of human society, from the emergence of a chief and a priest in a tribal group through princely states to modern industrial communities, can be visualised as a coherent progression. It involves the transfer of human activity from the provision of basic human needs for food, clothing and shelter to the satisfaction of material, aesthetic and spiritual urges, aspired to but unattainable in simple societies, and providing a fuller and more rewarding life in those able to achieve them. There are still societies of which the members spend almost all of their time meeting their basic needs, and at the other end of the scale there are the advanced communities in which the demand for food is met by the work of less than 5 per cent of the population. Our disillusion in recent years with the fruits of our development efforts should not blind us to the merits of the process, and the worth and human satisfaction that comes from liberation from the need to spend one's life keeping alive. Nor should we criticise those still at subsistence level if, in their impatience to better their condition, they seek to imitate a social system that we are now finding inadequate.

The sources of our disillusion are to be found in the incentives by which development is motivated, and their consequences for both individuals and communities. These incentives are of three main categories; personal wealth, personal power, and community welfare. The basic difference between the Western and the Communist approach to development lies in the difference in emphasis between the first two and the third of these incentives. The Western belief in individual liberty and the right to equality of individual opportunity has resulted in a society motivated by the incentives of individual wealth and power. These incentives have at all times been tempered by some sense of obligation to the rights and needs of the community, but such obligation has been regarded as an irksome, though necessary, restriction on the free interplay of market forces that leads to the advance of the community primarily through the advance of its individual members. In China and Russia the advancement of the community has come to be the dominant incentive to development. It is not the only incentive. Personal wealth has some influence and, in

33

Russia particularly, it appears that personal power still motivates at least some members of the community.

These incentives exist in all societies, and they are all important in the development process. They are all valid in their place, but it seems to me apparent that their relative value varies according to circumstances. Consider for example, the 'Green Revolution' situation in India. In the irrigated wheat region of the northwest, between Delhi and the Pakistan border, resources of land, fertiliser, water and credit have been available to the individual farmer. The opportunity to gain personal wealth has been open to the individual, and has proved to be a powerful incentive to improvement of productivity (see, for example, Aggarwal, 1973). The increase in wheat production has been such that wheat has risen from 12 per cent to 25 per cent of the food grain production of the whole of India. By contrast, in the rice areas there is much more interdependence between farmers, particularly in the management of water. Though community betterment is a more valid incentive than private gain, unfortunately it has proved a less powerful motive force than the sum of the personal incentives where individual gain is the prize. Hence, the productive potential of the new farming techniques has not been realised in the rice areas to the extent that it has in the irrigated wheat region.

Development under the stimulus of the hope of gain has served India very well where the circumstances were appropriate, but it is only valid when limited by other needs and obligations. The wheat revolution has been criticised by those who have observed its social consequences in terms of the polarisation of wealth: the tendency to concentrate resources in land, water and money in the hands of a few at the expense of the many. The failure to respond to the incentive of community betterment has been unfortunate in the prosperous wheat region just as it has been in the rice situation where a community response was the only one possible if there were to be any betterment at all. Thus, if development is to be effective, there must be a balance between the incentives of personal wealth, personal power and community betterment, and the balance will be different under different circumstances.

The dominance of the personal incentives in Western development derives in large measure from the circumstances of the Industrial Revolution, which favoured innovation by the individual for personal gain. Western economics is the economics of the entrepreneur, and the application of knowledge or the exploration of new fields by groups or societies has had little place in it. One may perhaps ascribe at least a part of the poor record of nationalised industry to the fact that progress in this situation calls for community action, in which we have little of either experience or inclination.

The incentive of community betterment has been much more important in Communist countries than in the West. An important factor in this has been the manifest failure of incentive to the individual to motivate worthwhile development in society. The American Friends Service Committee Delegation (1972) which visited China in May 1972 reported on the Chinese situation as follows:

But there seems little question that cohesiveness and and commonality have reached unprecedented levels in the past two decades. They contrast so sharply with the disorder, factionalism, divisiveness, and weak or absent public morality observable fifty or twenty-five years ago that one wonders how the change could have been accomplished at all, let alone in the 23 years since 1949. The best general answer is probably to be found, as the Chinese themselves would argue, in the cumulative degree of degeneration and weakness of the old society through the century or so which preceded 1949. Things had become so bad that most people were ready or even eager for radical and rapid changes and were willing to make heavy sacrifices to achieve these changes.

Nevertheless, Chinese history and tradition, which are very different from the history and traditions of the West, contributed to the establishment of the new code of practice that has had such a revolutionary effect on their society. The AFSC Delegation stated that:

> In China a central fact is that the totality of life of the average Chinese citizen is governed by group values. These values have deep roots in the past and have been nurtured and built upon by the government and the party under the leadership of Chairman Mao. One underlying principle seems to persist: order is still given the highest priority and is the guiding rule in organising the life of the country. The development of the group is more important than that of the individual.

For a community-oriented revolution, this historical background is an asset that the West does not possess. Moreover, the present climate of opinion in the West would preclude the kind of emphasis in education which the AFSC Delegation noted: 'the Chinese educational system inculcates values which emphasize group effort and self-sacrifice – not private development or individual success'.

Both the Chinese and the Russian revolutions were generated in circumstances that bordered on chaos, and in which the countries felt themselves threatened from without. These circumstances are very powerful inducements first to devising, and later to supporting, new systems of organising society. It is in periods of security, and of new and improved opportunity, that the prospects of individual gain and of personal power are most evident and most alluring. So the success of the Russian and Chinese revolutions may lead to conditions in which the inducements to the individual which motivated the Western Industrial Revolution may overtake the inducement of community benefit on which their revolutions have been built. On China, the AFSC Delegation commented :

> Consumerism is frowned on, and our efforts to discuss its insidious grasp on people as incomes and production rise, using the American example, were met with firm denials that such a problem could ever become a serious one in this society of the 'continuous revolution'. Nevertheless, we wonder about the future . . . Perhaps our Chinese friends are right that this will never happen to them; with an American example of excessive consumerism in our minds, we sincerely hope they are right, and wish them well.

In Russia, to the observer at a distance, it appears that the prospect of individual power is proving more attractive than the prospect of individual wealth, and that it is through the pursuit of power that the Russian revolution might change its character. For this the West must take at least some of the blame. The super-power strategy of balanced threat can hardly do other than encourage the pursuit of power by the individual as well as by the community.

Russia and China have established new social and economic systems. India, which is also poor and socially distressed, has been the location of social experiment but has not achieved the kind and magnitude of change of Russia and China. Beginning with Gandhi and continuing in various forms to the present day, there has been a succession of social experiments, all of which had for their motivation the denial of individual incentives and the pursuit of social betterment by way of community action. Gandhi set the pattern, with his emphasis on austerity, the amelioration of poverty by small-scale industry such as the manufacture of homespun cloth, and the social integration of the deprived, whether they were landless labourers or those excluded by the caste system.

Let it be quite clear that, though he achieved no social revolution such as was carried out by Chairman Mao, Gandhi has had an impact on Indian thought and practice that is profound and continuing. India does not practice what Gandhi preached, but his preaching has entered into her conscience, and has influenced her concept of the shape a just society would take.

It is pertinent to enquire why, if Mao could change the face of China, Gandhi did not reform India. It might be argued that it was because Mao put through his revolution by force, and this Gandhi was not prepared to do. Such an argument could not be sustained. India has repeatedly been overturned by force, by Indians and by invaders with a great diversity of religious and secular beliefs, and the outstanding feature of such revolutions is the closeness with which India after resembled India before. Revolution does not necessarily connote change, except in the minority at the top that holds the power.

In the tension between the individual and the community, India is ambivalent. There is no doubt that it is commoner in India than in the West to find men devoted to, and willing to make great personal sacrifices for, community betterment. There is also no doubt – witness the wheat revolution in northwest India – that where there is hope of gain and of personal advancement, Indian communities respond with an individualism that is of the kind that characterised the Western Industrial Revolution. This has resulted in a degree of polarisation. While the rich have got richer in the rich provinces, devoted and self-sacrificing Indians have shared, and endeavoured to ameliorate, the poverty of the poor, the Harijans and the tribals in the poor provinces. They have had their successes, but they have been small in relation to the size of the problem, and they show no sign of becoming self-propagating as the 'Green Revolution' became self-propagating in the northwest. Indeed, their successes have in large measure depended on the introduction of resources from outside. Government support, aid from overseas donors and the skills, intellect and commitment of workers such as those in the Gramdan movement have been necessary to get the village communities moving. Moreover, too often an advance made with aid given for a specific project leads, not to self-supporting growth and development, but to patient waiting for the next gift from without to energise the next move. This is partly because the villages involved have been chosen for their poverty, and in general actual poverty is associated with poor potential also. There are not the resources on which to generate a self-improving system without aid from outside.

A parallel to this experience is to be found in the rural development projects sponsored by the Shell International Petroleum Company (Cox *et al*. 1966). Their pioneer project at Borgo a Mozzano in Italy is a striking demonstration of the successful application of technical assistance to an isolated and poor peasant community. Their success has been achieved in a context of labour transfer out of agriculture, and of migration out of the area. Thus, some of the mountain land with the lowest potential has been abandoned altogether and the lower land offering greater resources is now exploited more effectively by a smaller farming population. In a sister project at Seva do Vouga in Portugal, achievement, though substantial, has been less striking because there have not been the same opportunities for transfer to other occupations, either locally or outside the district, and pressure on the limited local resources has not been relieved to the same extent.

Whatever the social motivation of those who try to alleviate poverty – the capitalism of Shell International or the non-violent, personal poverty and peaceful persuasion of the Gramdan movement – the hard facts of the availability of resources and the behaviour patterns of unregenerate man here and now determine the nature of the changes we achieve and the extent to which we approach a more just society. Erica Linton (1974), discussing a successful community betterment project in India, sums the matter up as follows:

We advocate that leadership be created in the village. Both at [Village P] and [Village B] strong men have emerged. Leadership by its very nature and function produces social inequality. The case of [Village P] also shows that, when men are not equal in the way they apply themselves to their tasks, *vide* the man who was too lazy to walk 100 yards to cultivate his plots, inequality is maintained, created or even aggravated.

In many cases the men to whom I spoke pleaded ignorance of what was going on but stated that their brother or father were attending the *gram sabha* meetings. Seeing both men together, the one who is involved and the one who is not, underlines the fact that there is inequality within the family, some members being more intelligent, willing and harder working than others. This not only applies to villagers but to families the world over. Add to this the use one family will make of added income from improved farming conditions compared with another – one spends it on consumer goods and drink, another in inputs to increase their yield further – inequality results in spite of the initial benefit having been the same.

'Leadership by its very nature and function produces social inequality', 'there is inequality within the family', and 'inequality results in spite of the initial benefit having been the same'. These are facts of life, in India, in China and in Britain, and it is in this context that men have sought, in different ways, to achieve a measure of social justice. We are not all equal, physically, intellectually, and in our inheritance from our own past. This is not to say that social justice is beyond us, but rather that it is not to be thought of in terms of equality. Diversity of abilities is among the great assets of mankind, and we need to conceive of social justice in the context of the exploitation of the qualities of leadership, of varying responses to opportunity, and of differing endowments from our historic past. We live with our past. We can watch the efforts of others and we can learn from them. We cannot profitably copy them, since their systems grow out of past circumstances that were not ours. Our own past attempts determine our present prospects, and it is by building on our past – on its failures as well as its successes – that we shall amend our present condition.

7

What kind of world do we want?

Human affairs are guided, not so much by ethical or spiritual principles as by codes of practice devised to provide a working relationship between those principles and the circumstances of the time. Circumstances change, and codes of practice are in fact modified and adjusted to meet those changes. In time, circumstances so change that it is no longer possible to meet new needs by further amendment, and progress falters and disillusion sets in until a new code is devised.

The present is such a time, and I want now to trace the history of recent codes of practice, and to lead to consideration of the criteria that must be met if we are to devise a new code to meet our current needs. What I have to offer is neither very new nor very original, but I believe it brings together and co-ordinates some rather fragmentary thought and debate about development in the Third World and progress in our own.

Our Western world is founded on a Christian heritage. This is not to say that it is a Christian world, but that it has been shaped by attempts over the last 2000 years to develop codes of practice that were moderated – however inadequately – by Christian principles. Two codes of practice, devised in the last few centuries, are particularly relevant to our present discussion. Up until the late eighteenth century, it was accepted in Britain, and indeed preached from the pulpits of English churches, that a man should be content with the station in life wherein he

43

was born. A similar, but more rigid, code of practice grew up in the Indian situation by the development of the caste system, and that only began to break up in the mid-twentieth century. In Britain and Western Europe, this code of practice disintegrated under the impact of the Industrial Revolution, and was replaced by the French Revolution cry of Liberty, Equality and Fraternity – liberty to do as you like within the law, equality before God and before the law, and brotherhood between men of common interests and aspirations. Both codes have much to commend them. There is much to be gained from a stable social pattern, in family life, in the kind of social relationships that foster neighbourliness and brotherliness, and in establishing an enduring spiritual dimension in society. But the economic opportunities, and the fundamental changes in social structure that arose in the late eighteenth and early nineteenth centuries, made advocacy of contentment with the station wherein you were born an intolerable constraint on the life of the individual. The new code, borrowed from the French Revolution, fitted admirably the needs of the British Industrial Revolution. Liberty meant liberty to take advantage of new and unprecedented opportunities, and equality meant equality of access to these opportunities, regardless of station in life. Fraternity came to mean the right to combine to further the ends of your own group.

Both these codes of practice were reasoned attempts to apply principles that were of Christian origin to the circumstances of their time. Just as the first became obsolete, and indeed opressive, through change in circumstance, the second has now become an obstacle to Christian practice. For in the last century liberty has commonly included liberty to take advantage of your neighbour, to climb over the back of the man in front, and to grasp all you can and hold it against all comers. And equality has at its best meant equal opportunities between unequal people so that the well-endowed outstrip their less fortunate fellows. It follows almost inevitably that fraternity is limited to the formation of interest groups, held together by the prospect of gaining power and advantage for their own association.

This code of practice – Liberty, Equality, Fraternity – has led us out of a static and rather poor society into a dynamic and wealthy one, and whatever its weaknesses and shortcomings we are – all of us – better off in material things that we should have been without it. I do believe, however, that it is now inadequate and indeed an obstacle to further progress. For since wealth generates wealth, this interpretation of liberty and equality leads to the concentration of wealth in the hands of a few – the rich get richer and the poor stay poor. We have followed the sayings of Jesus in the parable of the talents, but we have forgotten that the original capital was a loan and not a gift, and the increment and the power it gave were both to be surrendered when the lord returned.

We have hoped that a social system under which the total wealth of the community rose greatly would turn out to be one in which poverty disappeared. This has proved not to be the case. Not only is the increased wealth accumulated disproportionately by those already rich, but also there always remains a sector of society, albeit a small one, which is unable to take advantage of the new opportunities that the affluent society offers. There are the drop-outs of various kinds, some of whom have not the will, and others are without the

45

mental equipment, to cope with the complexities of modern society, and there are those who through misfortune, mistakes or mismanagement have lost – or have never possessed – the necessary basic assets without which it seems impossible to make a start towards a competence, if not a degree of affluence. These are the modern witness to the truth of Jesus' remark that 'the poor you have always with you'.

In our code of practice there is no solution to these problems of poverty, but the influence of Christian principles has been such that *ad hoc* attempts at solutions outside the code have been devised, and have had a mitigating effect. Even in the hey-day of the acquisitive economic system, there was always a strong current of philanthropy, exemplified by such enterprises as John Bellers' 'Colledge of Industry' that led to the beginnings of Quaker boarding schools, the private concern and political pressure that led to the Factory Acts, and the long campaign for the abolition of slavery. All these restricted the free play of the market economy.

In our own times, this piecemeal amendment of our economic system has gone much further, and much faster. Philanthropy in the medical field led to the development of a great range of voluntary hospitals, and these in turn formed a base for the development of the National Health Service, which is one of the great achievements of Britain in the mid-twentieth century. The Poor Law system and the Boards of Guardians of the days of my youth provided something of a refuge for the really poor. They were unpopular, and inevitably in the social climate of the time, the Workhouse carried a stigma that was to be avoided at all costs by the self-respecting. One of my early childhood memories is of my father's intense disapproval of a man in the village who was said to have allowed his parents to go into the Workhouse. Workhouses are now hospitals, or geriatric homes, and the much enlarged provision for social security is dispensed as far as possible in private homes.

None of this has been incorporated in a coherent socio-economic theory. These enterprises have been developed *ad hoc*, and we the people have given little thought to their impact on the code of practice by which we try to live our economic life. It is fashionable to criticise the 'profit motive' as the mainspring of our economy, but we have not thought out how we would pay for our social services without it. We are ready to meddle with the workings of the market economy when they affect our personal affairs. A large sector of the community – civil servants, welfare workers, the teaching profession, to name some of the most important – is rewarded with salaries derived from taxation, and we – and I myself fall into this class – have established what we regard as 'rights' to incremental scales based on age and length of service. Yet we have not worked out how to balance these increasing obligations with extra goods and services. We simply struggle to maintain our position, and we do not enquire whether, if we gain, some of our fellow men must lose.

The inadequacy of this code of practice, change and amend it as we may, has become the more apparent with the recognition that the resources available to us are limited. Conservationists have been concerned with this for some time, but it was only with the development of the energy crisis that the Western World really appreciated that the resources are not available to overcome poverty by unlimited economic growth. We must now recognise, on the one hand that for some of the world's most vital resources supply is no longer adequately responsive to demand, and on the other that society is no longer content to allocate resources simply on a basis of wealth. Thus the whole economic system of control by the price mechanism is in jeopardy. No rise in price will bring into the market resources that have been exhausted, and long before that situation arises, shortage and the accidents of monopoly lead to price rises that are disproportionate to any increases in supply that may be brought forward. Indeed, it is now clear that we can no longer rely on a price rise resulting in increased supplies at all. Producers who get a higher price may be content to produce less, since they will still get the money they need, and the conservation of their resources will be more important to them than their exploitation.

In theory, such a situation should result in the weaker buyers being squeezed out until the balance between supply and demand is restored. In practice, all the *ad hoc* amendments we have made in the market mechanism are designed to prevent it. This follows from the recognition of need in the allocation of resources. A plea for pensioners, for the sick and for the handicapped, carries great weight politically, and distributing at least some things according to need has become a part of our social thinking. Since those in need are by definition those least able to compete in the market, distribution according to need inactivates the mechanism for restricting demand to the available supply.

The controls of the old economic system have gone, but since demand must somehow be limited to the available supply, some new form of control must be devised.

The new control system must satisfy two criteria: allocation according to need, and a total allocation no greater than the supply available. Since it is apparent that supplies will be limited, the second criterion will only be met if those who now enjoy unrestricted access to supplies accept a restriction in their allocation. This is not an unfamiliar situation. It is a form of control that we in this country have in the past practised with success. It is rationing. Rationing is disliked and it is open to abuse, though no more so than the price mechanism. But if we really wish to distribute limited supplies according to need, rationing we must have, and we must learn to operate it as a control mechanism just as we have learnt to operate prices as a control mechanism.

Rationing is not to be entered upon lightly. It involves a substantial increase in the extent to which our activities are planned and controlled. Experience of the rationing systems operated in the second World War was that control of the size of the ration is not enough. The ration must be related to production and to price – for production must be paid for or it will cease. Indeed, the balancing of supply and demand in such a system can only be achieved by careful planning, which in turn will depend on adequate information on the supply situation and the orderly deployment of resources to make available sufficient to meet the need, as well as by the adjustment of the ration to conform to the limitations of supply.

Wartime rationing was almost exclusively of material goods – food, clothing, fuel, fertilisers and feeding stuffs – whereas today we are equally concerned with services. We demand a more comprehensive health service, more and better qualified social workers, higher standards and more generous student support in education. Thus there has now emerged a whole new area of interest in which we have committed ourselves, at least in principle, to allocation according to need. As we listen to – and endorse – pleas for more and better equipped doctors and nurses, social workers, and school and university teachers, we ought to ask ourselves where all these are coming from. In a situation of full, or nearly full, employment, the enlargement of social services can only come by the diversion of some people from other sectors of the economy. It is difficult to estimate the human resources that could be mobilised to meet the demands for further services, but the need to balance supply and demand is no less than with material goods. In the Third World, where educated and trained people are in short supply, manpower surveys have been made, to provide data on which to plan the expansion of services. They are difficult to make and are subject to the uncertainty that is always associated with anything so flexible as human capability. But when we clamour for increases in so many of the services that contribute to our affluence, it is surely sensible to make an assessment of the human resources on which they depend. Only on a basis of knowledge of the human resources actually and potentially available could we establish priorities in the development of social services, and implement them in so far as, and no further than, the available resources would permit. This would provide for services a counterpart to the rationing of goods. And it is no less necessary. It is no use demanding, for example, greater facilities for organ transplants and more provision for geriatric care or for abortions, without an assessment of the resources available for the purpose, and an allocation based on the evaluation of these against other social needs.

What kind of world do we want?

If we really intend to distribute goods and services according to need – which is surely the essence of any concept of social justice – we are committed to planning our economic life. For it is all too obvious that we cannot do everything, and if money is no longer to be the primary criterion in the assignment of priorities, the community must take precedence over the individual. If this is to happen, we shall need new incentives and new concepts of responsibility. Above all, we must escape from the delusion that we can command sufficient supplies to meet both market demand and human need without restriction. It is surely this delusion that has led us into inflation. Distribution according to need means that distribution according to wealth must be curtailed. Put bluntly, those of us who are rich must be content with less.

How much less, and how we calculate it, we have yet to work out. We must surely recognise that any further progress towards a just society will involve a conscious effort to set up a system of rationing for those goods and services that are to be distributed according to need, the establishment of a working relationship between that which is rationed and that which is distributed according to price, and the acceptance of sufficient discipline to ensure that we do not promise, either in the ration or in the market-place, more than we have to offer. We have recently had evidence of the aversion, almost the fear, of rationing in political and administrative circles. Petrol rationing was reluctantly organised, and thankfully abandoned when supplies eased. The demand from distributors for sugar rationing fell on deaf ears. This is understandable because rationing has been regarded as an administrative resort, devised to enable the community to survive abnormal conditions of stress. We ought now to regard it as an economic alternative, or at least supplement, to the price mechanism as the means of distributing resources, and as one having the merit of bringing the consideration of need into social calculations. So regarded, and so practised when opportunity arises – as with shortages of petrol and of sugar – it should become an acceptable addition to our community equipment.

Ours is a specialist society, and while by specialisation we have been able to achieve a level of affluence that an unspecialised community could not hope to reach, we have suffered the disability that to make any major change in the organisation of our affairs we must enlist the intervention of those central organs of government which alone can integrate the specialist sections of the community, and direct their activities. We recognise this in the way we invoke, or abuse, the all-powerful but faceless 'they'. 'They' ought to do something about it, we say, thereby disclaiming any power to do something about it ourselves. Yet 'they' are singularly incompetent in the face of the pressures of specialist interest groups. Like W. S. Gilbert's Duke of Plaza-Toro, they lead the regiment from behind. So while we recognise that in a specialist society we can only hope to control and direct the changes we desire through the central authority, yet the central authority will only be motivated to undertake such an enterprise if it is induced to do so by the activities of individual citizens. We must therefore consider what we, as individuals, can do to promote social justice in the community in which we live.

It is always easier to speak in general terms than to get down to the particulars of individual action, to talk about rationing as a code of practice rather than about restraint on consumption personally exercised. Indeed, when one is asked, as I have been, just what one should give up in planning to live a simpler life, one finds no quick or easy answers. But this is where a move towards greater social justice must begin, and I must at least set out my own thinking on the matter. I offer it, not as a panacea for the ills of maldistribution of resources, but as the way I have had to think out one facet of this problem, in which I have some expertise.

The challenge came to me early in 1974 when Stephen Sykes, then Dean of St John's College, asked me to produce for a Lenten message a brief statement on how we might 'eat more wisely: wisely, that is, both for our health and in relation to world resources'. It is not difficult to talk about world food supply prospects on the basis of the various statistical summaries that are readily available. To work out what we personally can do about it is a very different matter. But this surely is what we must do, and I offer my own thinking on a personal food policy as an example of the kind of thinking I believe we must do for ourselves, and the action that must follow from it, on a great range of resources – petrol and oil, other forms of energy, clothing and housing for example – that we have become accustomed to using without thought of the impact of our consumption on their availability to other people.

Cereals are the basic food supply for both man and his domestic animals. The current world food shortage is primarily a shortage of cereals, and any real amelioration of our present difficulties must come either from increasing the supply or from moderating the consumption of cereals. This is one of the resources of which the Western

53

World is a very heavy consumer. Lester Brown (1973) has set out the direct and indirect consumption of cereals *per caput* in selected countries. He summarises the situation as follows:

> In the poor countries the annual availability of grain per person averages only about 400 pounds per year. Nearly all of this small amount must be consumed directly to meet minimum energy needs. Little can be spared for conversion into animal protein.
>
> In the United States and Canada, per capita grain utilisation is currently approaching 1 ton per year. Of this total only about 150 pounds are consumed directly in the form of bread, pastries, and breakfast cereals. The remainder is consumed indirectly in the form of meat, milk, and eggs. The agricultural resources – land, water, fertiliser – required to support an average North American are nearly five times those of the average Indian, Nigerian or Colombian.

We in Britain follow not far behind the Americans in our pattern of food consumption, and we ought to remember that when we complain about the cost of food we are complaining about the cost of the diet of affluence.

We do not need so much of the expensive items in our diet; expensive, that is, in terms of the resources that go into their production. All of us who are over 20 years of age could cut down substantially our protein intake, and in particular our animal protein intake, without detriment to our nutritional standards. And, since in feeding people cereals go more than three times as far when eaten direct as they do when fed to livestock and eaten as meat, the potential contribution to world food supplies by cutting down meat consumption in the West is very great. We in Britain feed about $2\frac{1}{2}$ million tons of wheat to livestock every year. We could save a million tons of that for human food if we individually cut down our meat consumption, and in particular our consumption of the products of intensive livestock production – pig and poultry meat, and eggs – and we could do this ourselves, without administrative action by the powers that be. A million tons of wheat is not a large amount in world calculations, but a change in British eating habits of that magnitude would require a re-appraisal of agricultural and food supply policy by government, and so initiate the process of change that we so often regard as beyond our competence.

I set this out as an example of the analysis I believe we ought personally to make of the impact of affluence on our patterns of living. Simplicity of life style has a long history of commendation as a minor virtue. It has been drowned in the plenty of the last two decades. The plenty is ebbing away and simplicity may be forced upon us. Let us remember that it is a virtue. If we are to seek it, in our eating habits, in energy consumption, and in our personal expenditure, we must think out the distinction between our needs and our desires in the kind of way I have suggested for food.

I return to the question of what kind of a world we want. Our society at present is characterised by gross inequalities, and we have a long history of piecemeal amelioration of its worst injustices. The strong feelings we have that this is not enough, together with the realisation that there are not the readily available resources to provide everyone with the affluence we enjoy in the West, convince us that we must undertake more radical changes than we have hitherto contemplated. I suggest that we do not need – and indeed would not be able – to wipe the slate clean and start afresh with a new strategy. We have made substantial progress already in taking need into account in our social practice. We must now accept it more explicitly, and direct our thinking towards developing the meeting of need into a coherent code of practice. In material things, the time has already come to develop rationing as a means of ensuring just distribution of the necessities of life that are in short supply. In services, we have now reached the point where expansion to meet evident needs is very difficult to achieve, and here we must survey the situation, and settle in our own minds how much we can do with the human resources available to us. In a society plagued by labour shortages in some sectors of the economy and surpluses and unemployment in others, a manpower survey should lead on to retraining and redeployment, undertaken wholeheartedly and imaginatively. In our obsession with specialisation we have forgotten the flexibility and adaptability of the human mind, and we have been content to discard surplus people as we do obsolete equipment, when we could if we set our minds to it, re-equip them to exploit new opportunities.

If we are to progress towards this kind of society, we must begin by amending our own way of life. We live in such a specialised society that only centrally managed changes can serve our need, yet it is our experience that central authorities only move when induced to do so by the will of ordinary people. But, if we have the will, we can by our own actions create a situation in which action by central authorities must follow. Simpler living may be forced on us anyway, and I believe that if it is, we should meet it in a spirit of readiness and acceptance, and not with reluctance or opposition. But if not, if by some twist of economic fortune we are able to continue on the path we now follow, regardless of the injustices such a course involves, let us not excuse ourselves with the plea that there was nothing we could do. If we aspire to a greater measure of social justice, it lies with us to plan our life style to bring it about.

8

A beacon by which to steer a course

The argument of these lectures leads to the conclusion that those of us who are rich and powerful, either as individuals or as communities, should undertake some redistribution of our wealth and power for the benefit of those who, in present circumstances, are poor and powerless. Since the hope of wealth and power has been the mainspring of social development amongst us, I must, if I am to make my case, point to other incentives that I believe would motivate us to plan and carry out the sacrifices that are involved.

I began these lectures by reminding you of Eddington's conclusion that the essential difference between the physical world and the realm of spirit and mind 'seems to hang around the word "Ought" '. I have never got very far away from 'Ought' I have tried to show that though we have a well thought-out theory of the way our economic affairs work, we have nevertheless been impelled repeatedly, and increasingly as time went on, to meddle with its workings. We feel we *ought* to mitigate its effects on the poor and the vulnerable, and to restrict the benefits it bestows on the affluent and the powerful.

I hope I have made a convincing case for reconsidering the theory on which we work, with the prime objective of devising an effective instrument for the distribution of resources according to need. The incentive to undertake a reconsideration of our economic concepts can only be a conviction that we *ought* to do so. There are evil consequences of our current code of practice about which we have done something by mend-and-make-do expedients, and we now are convinced we must deal with them by a more radical reappraisal of our theories.

There have been times in our history when the concept of what we ought to do dominated thought and discussion, and indeed led to fearsome argument, disagreement, and even persecution. It is not so today, and I will give two examples of the detachment which characterises, and in considerable measure paralyses, thinking in the Western World. While I was Professor of Agriculture in Cambridge I sought from time to time the views of agriculturists, economists, and social scientists on what we ought to do about agricultural policy. We always ended by discussing what was going on at the time and what the consequences were likely to be of the policies we were then pursuing. We found no basis for discussion of what we ought to be doing. Similarly, I remember, shortly after the Belgian withdrawal from what was then the Congo, hearing a brilliant exposition of the situation by an academic who had just been there. We were greatly enlightened as to what was going on, but he, and we, were as uncommitted, and as free of any involvement, as if we were discussing a prize fight in New York or a football match in Glasgow.

Scientists have something to answer for in this climate of opinion. It is now academically respectable to evaluate a situation entirely impartially, so as to form an unbiased and comprehensive interpretation of it. So we have become much better at understanding events than at deciding how to try to influence them. I want to suggest that this is inadequate, and that it is more important to do something about a situation than it is to understand it.

Understanding helps in wise decision making, but decisions must be taken on such understanding as is available, and should not – indeed cannot – be postponed while evidence is collected or experiments are carried out. For deciding not to do anything yet is a decision – and more likely than not, a bad one.

This evidently brings us back to 'Ought', and raises the question of how we decide what we ought to do, and by what motives we are induced to try to do it. I have made my own enterprise more difficult by the line I have taken on past activities. I have explicitly disclaimed any intention to impute blame to courses of action that I have argued would be wrong in present circumstances. Indeed, I have emphasised the very substantial gains in human welfare that have resulted from the uncontrolled exercise of self-interest during the grand period of growth of the British economy, even though I have gone on to condemn such a system as intolerable now. So I am in danger of finding myself committed to the view that 'Ought' has no meaning, and right and wrong are concepts that can only be expressed in terms of expediency in the situation as it happens to be at the moment. Such a doctrine of expediency is very difficult to avoid unless one acknowledges some frame of reference, some spiritual beacons, against which a course of action can be checked. For such a frame of reference I look to the Christian faith. and in this approach I want to make it clear that the question I am asking is not 'should I believe' but the pragmatic question, 'do I get from it the guidance I need?'

I begin at the beginning. In setting out on his mission, Jesus began by withdrawing into the wilderness to consider what he should do. One must suppose that he later told his disciples of his thinking there, as this is the only way in which we could have heard what is embodied in the Gospels as the temptations of the devil. Looking at the sojourn in the wilderness as a time spent making up his mind on what he should do, three tempting choices offered themselves: first, to improve the well-being of his people by improving the farming of desert areas; secondly, to set up a military opposition to the Romans and conquer the world; or thirdly, to become a kind of religious magician and lead a fanatical religious movement. Jesus could have done any of these things, and this is the essence of temptation – the urge to do something we know we could achieve, but which we also believe is something we should forgo.

The Christian Gospel is a Gospel taught to a farming people living in a semi-arid land, and it is perhaps not surprising that the urban inhabitants of a moist temperate country such as ours have missed the point of the temptation to turn stones into bread. After six weeks' wandering in the wilderness, meeting the desert farmers, and living on what little he could get he must have been impressed both with the rigour of life in the desert, and with the potential in those areas for good land use. Recently, Evenari and his collaborators (1961) have given an account of the discovery and reconstruction of the ancient water harvesting system of farming in the Israeli desert. They have stated that the period 200 B.C. to A.D. 630 'represents the longest and most flourishing period of almost continuous settlement in the Negev'. This exploitation of the desert was what Jesus must have seen in his sojourn there, and reading the account by Evenari *et al.* of the techniques they have rediscovered by their study of the evidences of the old settlements, and the demonstration that in fact they work when re-

61

established, one can recognise the validity of Jesus' account of the episode as a temptation to turn stones into bread. If he had accepted this challenge, desert agriculture might have been extended, and might have survived longer before it fell into decay, but we should not have had the Christian Gospel. Similar considerations apply to the other temptations. The sojourn in the desert was the time when Jesus chose what he would do, and we have all that the Christian faith stands for because he chose to devote his life to people and not to a cause. When he died he left eleven men to carry on his work.

Jesus had his priorities right. In recent years we have been doing the job that he left as of secondary importance. The Israelis have redeveloped the Negev. India and Pakistan have created a green revolution on their irrigated lands. Australia has made the desert blossom, and in this country the unprecedented wheat yields of the 1974 harvest are evidence of the magnitude of our own agricultural revolution. All these great agricultural advances have been accompanied by social problems. The social unrest exemplified by landlord and tenant relationships in India, the Australian economic dictum 'get big or get out', and Britain's 'small farm problem' are symptoms of the inadequacy of our attempts to solve human problems through economic and social systems. These systems we need, but only if we have a spiritual system can they be made to work. Priority considerations are: 'But I say unto you that ye resist not evil: but whosoever shall smite thee on thy right cheek, turn to him the other also' and 'Give to him that asketh thee, and from him that would borrow of thee, turn not thou away.'

If we are honest, we shall admit that we do not

believe it would work, that you could not run human communities that way, and so we do not try. So Christianity has never yet won the battle against worldly wisdom. Jesus was crucified and the fears and self-interest that led him to the cross have dominated society ever since. Yet in defeat the Christian ideal is absorbed, taken over by stealth so that the material world that is still dominant is continuously modified by the faith we suppress.

I come back to Eddington's 'Ought'. This sense of obligation, this feeling that there are attitudes and actions required of us that go right beyond our individual interests, was the spiritual urge that Jesus set out to activate in the men he met and talked to. In recent times we have gone the other way. The individual has been at the centre of our thinking, and his development and his personal achievement have been major aims of our society. In a curious way this is a perversion of the Christian care for one another. We began by arguing that we must respect the rights and aspirations of others, and have ended by believing that the rights and aspirations of individuals – including ourselves – override all else. So liberty came to mean the liberty of the individual and not to connote responsibility, equality to mean on the one hand equal opportunity to profit at another's expense, and on the other, unashamed covetousness. And fraternity for most of us means little more than general benevolence greatly eased by a social system that ensures that most of our brotherly contacts are with people like ourselves.

These are the guiding conventions of our society, and the concentration of wealth and the exercise of power with which we are so much concerned goes on under these conventions. Neither great wealth nor great power are evil in themselves. They become evil when exercised for selfish ends, and selfish ends have been our motive force. It is more important to change the goals of our society than to alter the distribution of wealth within it. I naturally look to the founder of my faith for inspiration, but I, and I suspect most other people, have met in my life men who have had those objectives, founded in the spirit of man, that give honour and purpose to wealth and power. I will name two. Joseph Rowntree, wealthy manufacturer, devoted himself in his lifetime to the deployment of his wealth for the improvement of social conditions, and ensured the continuation of his concerns by the endowment of the Rowntree trusts. Carey Francis, Fellow of Peterhouse, Cambridge, gave up his university career to teach in mission schools in Kenya, and was one of the foremost of those who ensured that the security forces withstood the temptation to match terror with terror in the Mau Mau emergency. Joseph Rowntree never appeared to be wealthy, and Carey Francis would have been astonished if he had been accused of being powerful. Their interests and their commitment were unselfish, but they nevertheless exercised wealth and power.

It is this kind of spiritual exercise that we need. I do not find it easy to put into words my apprehension of the deepest things in man's nature, but I must try. In what I call the exercise of the spirit there is something of reverence and of worship, something of hope and expectation, and more than a little of accountability. We do not live our lives to ourselves. We have the expectation on the one hand of participation in the future, and on the other of accountability for the past. It is this that gives us the sense of the eternal worth of our actions, and is the ground of that sense of 'Ought' by which we transcend the motive of self-interest.

I have said that I look to the founder of my faith for inspiration. Jesus set out 2000 years ago the principles of hope and of accountability that should govern the exercise of wealth and power. His principles have become like a pearl in an oyster. Their irritation to our consciences has led us to overlay them with the lovely coating of our charitable activities. Lovely indeed, but what were meant to be the guiding principles of our lives have been insulated from our daily activities. The acquisition of wealth and power in response to self-interest has been protected from the irritation of feelings of community responsibility by the allocation of part of the wealth to social concerns.

These beacons have stood for 2000 years, and we have still made only modest progress by their light. It would be unrealistic to suppose that by drawing attention to them again, there is a prospect that we shall now move much more rapidly. Let me turn again to the temptations. If Jesus had set out to change the world quickly, he would either have chosen the role of the religious fanatic or have embarked on a career of conquest. And the changes he made would have been ephemeral. So with us also. Revolution is no solution. The just society will come only as we conceive it and bring it to birth, individually and collectively. It will not come by the bulldozer technique, by clearing the site and starting afresh, but by amendment and improvement, by building on what we have, by removing what is obsolete and mistaken. For a human community is not an architectural conception but a living thing, and continuity is of the essence of life.

65

I offer no quick solution to our problems. I only offer the old and untried one of concern for our fellow men rooted in a spiritual conviction. I myself draw from the Christian fountain, but there are others also, notably the Gandhian teaching in India. Self-interest was never an adequate incentive. In a modern society, with its intensity and complexity, it is patently a menace to our survival. The alternative of care for the community is as old as self-interest. We need to try it out.

John Drinkwater set the matter out in his

PRAYER

We know the paths wherein our feet should press
    Across our hearts are written thy decrees
Yet now, O Lord, be merciful to bless
    With more than these.

Grant us the will to fashion as we feel
    Grant us the strength to labour as we know,
Grant us the purpose, ribbed and edged with steel
    To strike the blow.

Knowledge we ask not – knowledge thou hast lent
    But Lord the will – there lies our bitter need
Give us to build above the deep intent
    The deed, the deed.

Let me return to the Third World. Our way of life, with its self-interest and its gross inequalities, offers them no more than the prospect of an affluent minority in a sea of poverty, and with that neither they nor we will be content. The challenge of the Third World is that it has shown us convincingly that we must do something about our own.

References

Aggarwal, Partap C. (1973) *The Green Revolution and Rural Labour.* Shri Ram Centre for Industrial Relations and Human Resources, New Delhi.

American Friends Service Committee Delegation (1972) *Experience Without Precedent, Some Quaker Observations on China Today.* AFSC, Philadelphia.

Blackett, P. M. S. (1967) 'The ever-widening gap', *Science* vol. 155, pp. 959–64.

Brown, Lester R. (1973) *Population and Affluence: Growing Pressures on World Food Resources.* Development Paper 15, Overseas Development Council, Washington D.C.

Cox, I. H., Virone, L. E., Pellizzi, C., Upton, M. and Marcano, L. (1966) *The Transformation of Rural Communities.* British Association for the Advancement of Science, London.

Eddington, A. S. (1929) *Science and the Unseen World,* Swarthmore Lecture. George Allen & Unwin, London.

Ernle, Lord (1919) *English Farming, Past and Present.* Longmans, Green & Co., London.

Evenari, M., Shanan, R., Tadmor, N. and Aharoni, Y. (1961) 'Ancient agriculture in the Negev', *Science* vol. 133, pp. 979–96.

Linton, Erica (1974) 'Amarpurkashi and Agrindus' (ms.).